# Clean Technology

## A Research Handbook for a Green Society

**Koso Brown**

# Contents

# Introduction

The world is changing due to technology, and this trend will only intensify. Whether the shift has been for the better or worse is usually up for debate. Clean Tech can help with this.

The goal of the field of "clean technology" is to lessen human effects on the environment.

Because governments all over the world have declared a climate emergency, there is a push to adopt clean technology solutions. This is a direct outcome of our energy dependence since the Industrial Revolution on fossil fuels. Governments are enacting laws requiring businesses and people to lessen their environmental effects.

Schemes for voluntary reduction and carbon trading are being introduced to support these regimes. There has to be a broad adoption of clean technology both now and in the future as a result of the increasing pressure from international communities and organizations.

# Chapter 1

# What is meant by "clean technology?"

First, let's define clean technology. What exactly does the term "clean" mean? In brief, clean technology refers to any action done to mitigate, or better yet, completely eradicate, adverse environmental effects while promoting social and economic advancement.

Clean technology revolves around using resources that are non-renewable and acting to protect them. Natural resources should be conserved or avoided if possible.

Clean technology seeks to increase efficiency and output while getting rid of or minimizing waste and pollution. Every stage of the life cycle of a product or process should see the implementation of these modifications.

## Key Clean Technology Sectors

The following are now the clean technology industry's major sectors:

- ❖ Waste management and recycling
- ❖ Reduction of pollution
- ❖ The supply of pure water to everyone who needs it
- ❖ Energy optimization and sustainability to lessen reliance on fossil fuels

## Waste management and recycling

- ✓ **Recycling of consumer products:** The vast majority of consumer goods on the market today have numerous recyclable parts or components. A product should be properly classified once it reaches the "post-consumer" stage to prevent it from ending up in landfills.

- ✓ **Reduction and treatment of toxic waste:** Toxic or hazardous waste, which frequently takes the form of materials or chemicals that are dangerous, needs to be handled cautiously. Plans for the reduction, collection, treatment, and regulation of hazardous and toxic waste are in place in the majority of countries.

## Reduction of pollution

✓ **Cleaning up contaminated areas:** Landowners must act quickly to ensure that an area is rendered safe for people and animals upon receiving a request for environmental cleanup from the government or other land remediation authority. This can entail eliminating toxins and pollutants from sediment, surface water, groundwater, and soil.

✓ **Monitoring of pollutants:** The State Local Air Monitoring Stations (SLAMS) system, which uses air monitoring stations across the country, generates yearly statistics on the amount of pollutants present in each state. Throughout the world, similar methods are used to track the amounts of air pollutants, such as using drones and satellites to produce unbiased, accurate monitoring data.

✓ **Control of emissions:** Here are a few strategies to lower emissions worldwide. Some recommended methods of reducing emissions include switching from conventional gasoline-powered automobiles to electric and biofuel-powered vehicles, as well as implementing clean technology like the sustainable energy sources mentioned above.

**The supply of pure water to everyone who needs it**

- ✓ **Water treatment:** To make raw water suitable for human consumption, it must be treated.
- ✓ **Wastewater treatment:** Wastewater treatment is the process of turning it into water so that it can be recycled or added back into the water cycle.

**Energy optimization and sustainability to lessen reliance on fossil fuels**

Numerous technologies minimize energy consumption or harness renewable energy sources to lessen reliance on fossil fuels.

✓ **Energy reduction:** This includes automated systems, human behavior management, sustainable development, and sustainable building management—all methods of reducing energy use from the demand side.

✓ **Geothermal energy:** This is just the earth's natural heat. Similar to how focused light beams are used in CSPs to heat water to drive heat engines and produce electricity, this heat can also be employed by building systems to harvest the naturally occurring thermal energy in the earth for space heating.

✓ **Smart energy:** This refers to the various ways that the introduction of automated energy distribution, responsive energy supply, and connected energy consumption can optimize energy usage; these developments are made feasible by the emergence of the Internet of Things (IoT).

✓ **Wind power:** Large wind farms, which are frequently located offshore, are the usual method of utilizing this renewable resource. A wind farm is made up of many separate wind turbines that are connected and produce power without emitting greenhouse gases during the construction process.

✓ **Hydroelectric Generating Energy:** This is the process of creating electricity by harnessing the gravitational force of falling or flowing water. When compared to methods based on fossil fuels, a hydroelectric power plant will generate incredibly minimal amounts of greenhouse gases once it is built.

✓ **Solar energy:** There are two methods for using solar energy to create electricity: photovoltaics (PV) and concentrated solar power systems (CSPs). While the latter concentrates a light beam onto a tiny region using lenses or mirrors, the former exploits the photovoltaic effect to directly convert light into an electrical current. This is transformed into heat, which powers a heat engine to produce energy.

# Chapter 2 Clean technology examples

Many businesses are actively attempting to become involved in sustainable technologies in the current environment. There are numerous methods for lowering your carbon footprint right from the start. Let's examine a few instances of the application of clean technology.

## Hydroponics

Given the significant role that agriculture plays in global warming, we must come up with a solution and begin applying smart agriculture practices. One method for achieving this is hydroponics. Using nutrient-enriched water to grow crops instead of soil is an example of this kind of eco-innovation.

Hydroponic farming has numerous advantages over conventional farming, including decreased space requirements, crop mobility, less water usage, and no seasonal restrictions. Nevertheless, because this agricultural method is far more costly than traditional farming, not everyone can afford to switch.

**Sustainable transport**

A significant amount of greenhouse gas emissions, which eventually cause global warming, are caused by transportation. Transportation accounted for 29% of all greenhouse gas emissions in the US in 2019. For work, school, social events, and other purposes, we utilize transportation.

It's unsettling to see how much of an impact transportation has on the environment for something we use so frequently. Fortunately, there are more environmentally friendly transportation options becoming available; electric cars, for example, are now widely accessible and in use.

Adding more public transit is an excellent method to incorporate clean technology, as are electric substitutes for fossil fuel-powered vehicles. Less personal automobiles are adding to gas emissions as more people use shared transportation.

## Material recycling

Despite their apparent simplicity, recycling, and upcycling are incredibly effective ways to boost sustainability initiatives. Recycling is a game-changer for sustainability, whether it takes the form of people reusing bottles to cut down on plastic waste or businesses repurposing trash to create new products.

Particularly in the area of fashion, sustainability is growing as more people become conscious of the negative effects of rapid fashion. Numerous fashion brands, like Levi's, implement clothing recycling programs to repurpose used clothing into new items. Incentives and discounts are frequently provided by these brands to encourage customers to return any unwanted clothing.

**Several elements influencing clean technology**

- ➢ **Food waste:** This is food that is thrown away even though it is suitable for human consumption. We now have a significant problem with 1.3 billion tons of food wasted annually. Food is wasted at every level of the production, processing, distribution, and even consumption processes.

- ➢ **Biodiversity loss:** The diversity of life on Earth is known as biodiversity. The loss of biodiversity has a major impact on the health of the planet. Because all living things depend on other species to survive, the loss of biodiversity can have disastrous cascading effects.

➤ **Pollution:** Pollution can be defined as the introduction of a toxic material into an unspoiled area. Air pollution, water pollution, and land pollution are three types of pollution that can be very dangerous to life.

➤ **Climate change:** Climate change, which is arguably the most well-known environmental issue, is the alteration of a region's average temperature or weather patterns. The average global temperature is rising alarmingly, and it will only get higher if we don't take action to lessen our carbon impact.

**How Clean Technology Can Bring About Change**

➤ **Water quality:** Water quality is a major environmental problem that is frequently disregarded. Numerous businesses are working to cut down on water use and supply undeveloped areas with clean water. Water treatment and wastewater treatment are two areas where clean technology can be used with water.

➤ **Sustainable energy:** Many clean tech techniques aim to produce and use energy sustainably. The world continuously uses electricity, and technological advancements like solar energy, wind turbines, and geothermal energy can have a significant impact.

➤ **Waste reduction:** Reducing the quantity of rubbish we send to landfills is crucial since space on our planet is rapidly running out. Any initiative to lessen waste generation or even recycling can benefit from the use of clean technologies.

# Chapter 3 Who benefits from clean technology?

In the end, the adoption of clean technology benefits all of us; every advancement toward sustainable technology contributes to environmental preservation. Though it might be difficult to see how you or your business can improve the state of the planet, every little bit of sustainability count.

Clean technology can be used in almost any sector. By using clean technology, even startups and small businesses may have an influence. Even if certain sectors harm the environment far more than others, we can all do our share to protect the world's natural resources. Energy, waste management, and agriculture are a few of the industries that have the worst effects on the environment. As we approach Earth's tipping point, however, it will be too late to save the ecosystem if all industries do not act.

The concepts of corporate social responsibility and sustainability are growing trends that motivate companies to improve the environment.

## The demand for clean technology

More needs to be done to promote sustainability worldwide as environmental problems and population growth coincide. Our natural environment is changing dramatically, therefore we must act now before things get worse.

The Earth's average temperature is rising, there is a growing shortage of water, and an increasing number of animals are going extinct. Unfortunately, human activity is mostly to blame for these bad developments, so we must adopt sustainable technology to lessen these damaging behaviors.

## The sustainable generation

A generational shift in our society may be the cause of the increased awareness of sustainability. Today's younger generations are more likely to make sustainable decisions and to be more environmentally sensitive.

According to reports, 73% of Generation Z is willing to pay more for sustainable items. Gen Z customers are more inclined to base their purchases on a company's environmental and ethical policies. Future clean technology should be anticipated given the younger generations' strong interest in protecting the environment.

**Trends in clean technologies**

Nowadays, practically every business plan and operation prioritize what was formerly an afterthought when it comes to running a business. Let's examine some of the ways that clean technology trends are likely to keep expanding in the future.

- **Smart agriculture**

As we just touched on, smart agriculture is revolutionary in terms of lowering carbon emissions. Significant environmental effects result from the agriculture sector; just the livestock business accounts for 44% of methane emissions caused by humans.

A small portion of the field of smart agriculture includes hydroponics. With the application of technologies like artificial intelligence and precision farming in the agricultural industry, there is much to explore in the field of technology.

- **Impact investing**

Investing in businesses that have a beneficial environmental impact in addition to making money is known as impact investing. These impact investments are directed toward businesses whose primary business strategy is to continuously improve.

Businesses contemplating a clean-tech strategy are further incentivized by the growth of impact investing. Cleaner technology is a driving force behind the global shift towards a more sustainable lifestyle.

Given the speed at which the clean technology sector is expanding, it makes sense for investors to be focused in this area. Impact investment also grows as cleaner technology advances; in 2020, the market was projected to be valued at 715 billion USD.

- **COP26 impacts**

Even though COP26 isn't yet finished, it's already having an impact on national climate change commitments and strategies. At the event, global leaders will discuss a wide range of topics, including gender, money, fashion, and education.

After the conference is done, we should anticipate more advancements to be put into practice, but here are some of the outcomes from COP26 thus far:

- ❖ By 2030, India wants to have zero net emissions.
- ❖ Africa is getting ready to invest USD 6 billion in adaptation to meet the demands of climate change.

- ❖ More than 20 nations and financial organizations have committed to stop funding any overseas fossil fuel development projects.
- ❖ By 2030, US President Joe Biden wants to cut methane emissions worldwide by thirty percent.

- **Artificial intelligence**

Machine learning is used by this sophisticated computer program to mimic human intellect. Artificial intelligence finds application not only in agriculture but in many other areas as well. AI can complete jobs with extreme precision, which reduces a considerable deal of waste in terms of time or resources.

Because AI can automate numerous production chain steps, it has shown to be especially helpful in manufacturing processes. Moreover, it can save expenses, which can then be allocated to more environmentally friendly technology.

# Chapter 4 How Can Clean Technology Reduce Global Warming?

Fossil fuels have been used by humans for more than 150 years, and as their use has grown, so has the emission of greenhouse gases during their combustion. Earth's temperature rises as a result of these greenhouse gases' ability to trap heat in the atmosphere. Rising sea levels, more frequent extreme weather events, shifting wildlife habitats and populations, and other effects are all signs of climate change, of which global warming is one symptom.

Renewable-produced energy does not contribute to global warming because it does not release greenhouse gases like carbon dioxide. Because of these renewable energy sources, climate change is not progressing, and actions like reforestation can lessen the harm already done to the planet, hence lowering global warming.

# Can Fossil Fuels Be Replaced by Clean Technology?

Since fossil fuels have been used by humans for many years—decades—the transition to renewable energy has been relatively recent. Because of this, people still view renewable energy sources as unpredictable and unable to supply the world's electricity needs. This implies that carbon-based energy sources continue to supplement renewable energy.

However, it's thought that by effectively storing renewable energy, we can balance our energy needs and use it when needed. Many efforts are being made to enhance the clean energy infrastructure and storage capacity; according to experts, by 2050, clean renewable energy may displace fossil fuels.

# Chapter 5 How Will Clean Technology Help Our Economy?

There are financial advantages to clean energy, not the least of which stem from the labor created to manufacture and install clean energy solutions, as well as to upgrade the infrastructure. As the globe starts to move away from fossil fuels, the renewable and clean energy industries are growing, which means that additional opportunities will exist in everything from power generation and storage to eMobility.

The knowledge that comes with creating these next-generation power solutions can help those who acquire it, providing contracts and jobs to those who adopt renewable energy more slowly.

The true motivation for clean energy is, of course, improving the planet's future; the financial consequences are but one aspect of the story. However, as the usage of fossil fuels decreases, so will the corresponding financial benefits; therefore, clean energy is not only beneficial to the environment but also a step forward for the industry

## How Can Clean Technology Be Obtained?

❖ In order to generate power, biomass employs solid fuel made from plant components. This energy source is not wood; instead, it is far cleaner and more energy-efficient than it was in the past, even though it still involves the burning of organic materials. It is not only cost-effective but also environmentally beneficial to use household, industrial, and agricultural waste as solid, liquid, or gas fuel.

❖ The planet's most plentiful and freely available energy source is sunlight; in just one hour, enough solar energy reaches Earth to meet the planet's entire annual energy needs. The time of day, the seasons, and one's geographic location all affect solar power, of course. Despite this, there are now large and residential uses for solar energy.

❖ One of the renewable energy sources with the greatest commercial development is hydropower or water power. In addition to being easier to store produced energy so that it can be used by demand, this energy source is said to be more dependable than solar or wind power. Additionally, municipal hydropower is being researched, which means that one day all of us will be able to produce electricity in our houses by letting the water run through pipes. Large-scale hydropower is derived from tidal power, which is extremely predictable and dependable while not offering a steady stream of energy.

❖ Another abundant sustainable energy source is wind power, with wind farms making a significant contribution to the UK's and other countries' electricity supplies. Although it is currently possible to obtain household "off-grid" wind energy, not every property is suited for a wind turbine.

## Is Clean Technology Clean?

By definition, all forms of energy are "clean," but not all forms of renewable energy are as well. Burning wood from forests that are maintained sustainably, for instance, can be renewable but is not clean because it releases carbon dioxide into the environment.

Sources like solar and wind energy are regarded as fully clean and renewable because they have zero carbon costs associated with generation and storage

# Conclusion

We hope that after reading this book, you will have a solid understanding of clean technology and what the future holds for the sector. With any luck, clean technology will only become more and more important as we move toward a more sustainable future.

When our reliance on fossil fuels declines, clean energy seems to be the way of the future for the world's power demands. The cost of developing and implementing these new power solutions will decrease as the push towards clean, green, and renewable energy continues.

The benefits of clean energy for the environment, society, and economy are becoming more widely acknowledged, and this trend will continue as more countries, states, and cities join the green power agenda.